I0503297

CONTENTS

LIST OF ACRONYMS

ANL	DOE's Argonne National Laboratory
BioPA	biological performance assessment
BPA	Bonneville Power Administration
BSOA	Basin Scale Opportunity Assessment
CH	conventional hydropower
Corps	U.S. Army Corps of Engineers
DOE	U.S. Department of Energy
DOI	U.S. Department of the Interior
EERE	DOE's Office of Energy Efficiency and Renewable Energy
EPA	U.S. Environmental Protection Agency
FCRPS	Federal Columbia River Power System
FERC	Federal Energy Regulatory Commission
FIHWG	Federal Inland Hydropower Working Group
FOA	funding opportunity announcement
FWS	Fish and Wildlife Service
FY	fiscal year
HAC	Hydropower Analysis Center
HDC	Hydroelectric Design Center
HFM	hydropower flow measurement
HK	hydrokinetic
HMI	Hydropower Modernization Initiative
MOU	Memorandum of Understanding
NHA	National Hydropower Association
NHAAP	National Hydropower Asset Assessment Program
NOAA	National Oceanic and Atmospheric Administration
NREL	DOE's National Renewable Energy Laboratory
O&M	operations and maintenance
ORNL	DOE's Oak Ridge National Laboratory
PMAs	power marketing administrations
PNNL	DOE's Pacific Northwest National Laboratory
PSH	pumped-storage hydropower
RAPID	Regulatory and Permitting Information Desktop
R&D	research and development
Reclamation	DOI's Bureau of Reclamation
SEPA	Southeast Power Administration
SNL	DOE's Sandia National Laboratories

SWPA	Southwest Power Administration
TDG	total dissolved gas
TSC	Technical Service Center
TVA	Tennessee Valley Authority
USFS	U.S. Forest Service
USGS	U.S. Geological Survey
WAPA	Western Area Power Administration
WWCRA	West-Wide Climate Risk Assessment
WUOT	Water Use Optimization Toolset
WWPTO	DOE's Wind and Water Power Technologies Office

EXECUTIVE SUMMARY

Since the U.S. Department of the Army (through the U.S. Army Corps of Engineers), U.S. Department of Energy (through the Office of Energy Efficiency and Renewable Energy), and U.S. Department of the Interior (through the Bureau of Reclamation) signed their Memorandum of Understanding (MOU) for Hydropower in 2010, these agencies have endeavored to advance their mutual goals for greater development and utilization of clean, reliable, cost-effective, and sustainable hydropower generation in the U.S. This Action Plan renews the commitments of the three agencies to the 2010 MOU and details the second phase of collaboration, which seeks to support the Administration's goals for doubling renewable energy generation by 2020 and improving federal permitting processes for clean energy as established in the President's Climate Action Plan. Through continued collaboration and partnerships with other federal agencies, the hydropower industry, the research community, and numerous stakeholders, these agencies intend to continue working toward the objectives and goals of the MOU.

The MOU agencies have complementary and overlapping roles in the domain of hydropower research, planning, development, and operations. This Action Plan provides a structure for collaborative activities that clarifies those roles and enhances the efficiency and benefits of coordinated activities in the following areas: (a) Technology Development, (b) Asset Management, (c) Hydropower Sustainability, (d) Quantifying Hydropower Capabilities and Value in Power Systems, and (e) Information Sharing, Coordination, and Strategic Planning. The sections that follow describe the roles of the MOU agencies in each of these areas and the specific action items on which the MOU agencies plan to collaborate to accomplish their goals.

If successful, these collaborations could help to:

- Improve the accuracy and reduce costs of flow measurement technology, which, if successful, could increase generation at existing plants and improve productivity of new hydropower systems yet to be installed.
- Establish testing protocols and analysis tools to aid developers and owners/operators of irrigation canals to more rapidly evaluate hundreds of new hydrokinetic projects across the country.
- Evaluate new superconducting generator technology, which could significantly reduce size and weight of generators for new hydropower projects, leading to reduced civil works costs, and could potentially reduce losses and increase generator output for existing facilities.
- Further develop low-impact, low-cost hydropower technologies suitable for demonstration and deployment at non-powered dams and conduits, where there is potential to increase U.S. hydropower generation by more than 17,500 gigawatt hours per year (equivalent to powering over 1.5 million U.S. homes per year).[1]
- Develop design tools to improve the environmental performance of hydraulic turbines for replacement and new development applications, reducing turbine-passage injury and mortality rates for sensitive aquatic species.
- Assess the risks to U.S. hydropower generation and water infrastructure from climate change.

Additionally, the MOU agencies remain committed to focusing attention and effort on improving regulatory processes for hydropower across the nation. The Bureau of Reclamation plans to implement the Bureau of Reclamation Small Conduit Hydropower Development and Rural Jobs Act (H.R. 678), and each of the MOU agencies intend to work collaboratively to assist the Federal Energy Regulatory Commission in implementing the Hydropower Regulatory Efficiency Act of 2013 (H.R. 267). In the near term, the U.S. Army Corps of Engineers and the Department of Energy intend to work collaboratively with the Federal Energy Regulatory Commission to improve regulatory efficiency for private and municipal development and operation of power facilities at the U.S. Army Corps of Engineers non-powered dams. Beyond these near-term efforts, the Department of Energy will lead projects (coordinating with MOU partners where appropriate) to promote standardization and modularization of hydropower components and systems targeted for deployment at non-powered dams and in new stream-reaches, while constantly seeking to improve the environmental performance of new technologies.

---

[1] Estimate based on resource assessments from the U.S. Army Corps of Engineers (http://cdm16021.contentdm.oclc.org/cdm/ref/collection/p266001coll1/id/2333) and the Bureau of Reclamation (http://www.usbr.gov/power/AssessmentReport/index.html and http://www.usbr.gov/power/CanalReport/index.html)

INTRODUCTION

As the largest source of renewable electricity generation in the United States, hydropower provides a wide range of benefits to the country. Hydropower is a minimal-emission, low-cost source of energy that can be relied upon for sustained and secure production of domestic electricity. Hydropower also provides consistent, reliable generation that can be quickly adjusted and dispatched to meet the electric grid's various needs.

On March 24, 2010, the Department of the Army—implemented through the U.S. Army Corps of Engineers (Corps)—the U.S. Department of Energy (DOE)—implemented through the Office of Energy Efficiency and Renewable Energy (EERE)—and the U.S. Department of the Interior (DOI)—implemented through the Bureau of Reclamation (Reclamation)—signed the Memorandum of Understanding (MOU) for Hydropower.[2] The purpose of the MOU is to "help meet the nation's needs for reliable, affordable, and environmentally sustainable hydropower by building a long-term working relationship, prioritizing similar goals, and aligning ongoing and future renewable energy development efforts." Additionally, the MOU aims to "(1) support the maintenance and sustainable optimization of existing federal and non-federal hydropower projects, (2) elevate the goal of increased hydropower generation as a priority of each agency to the extent permitted by their respective statutory authorities, (3) promote energy efficiency, and (4) ensure that new hydropower generation is implemented in a sustainable manner."

Since the signing of the MOU, the three agencies have sought to accomplish these goals, and intend to continue working together to do so. In April 2012, the agencies released a two-year update on their activities under the MOU that have occurred since its initiation.[3] In October 2012, the agencies met in Portland, Oregon, to review coordination for ongoing activities and discuss ideas for new collaborative efforts. This new Action Plan constitutes a recommitment of the agencies to the goals and principals of the MOU and details the joint activities that the agencies intend to work towards in the years to come.

The MOU agencies have complementary and overlapping roles in the domain of hydropower research, planning, development, and operations. This Action Plan is structured to provide clarity of collaborative roles that enhances the efficiency and benefits of coordinated activities in multiple areas:

- **Technology Development**: DOE invests federal resources in research and development (R&D) activities that will reduce costs and promote the deployment of new technologies in the long-term. While some DOE investments may have energy technology deployment payoffs that are immediate, others will come to fruition during hydropower development that is more than a decade in the future. Corps and Reclamation make mission-specific and site-specific hydropower technology R&D investments that typically have more immediate benefits to their own assets. The impact of MOU collaboration in this sub-domain is that multi-agency coordination of mission-specific, site-specific, and near-term technology R&D investments and activities can be enhanced for greater returns, and those investments can be leveraged within longer-term energy technology development strategies advanced by DOE.

---

[2] The signed MOU can be accessed at: http://energy.gov/eere/water/downloads/hydropower-memorandum-understanding.

[3] The report can be accessed at: http://energy.gov/eere/water/downloads/hydropower-memorandum-understanding

- **Asset Management**: Asset management is the systematic process of deploying, operating, maintaining, and upgrading assets cost-effectively, in a prioritized way, and with limited resources. In the context of this MOU Action Plan, assets are water control projects and components, including all equipment, structures, and water conveyances and reservoirs residing within the project boundaries. Assets may also include the sensors and control systems that link physical projects to centralized dispatch facilities. Corps and Reclamation are the primary managers for federal hydropower assets, with support and input from the four power marketing administrations (Bonneville Power Administration [BPA], Western Area Power Administration [WAPA], Southwestern Power Administration [SWPA], and Southeastern Power Administration [SEPA]).[4] These agencies must make asset management decisions amidst technical uncertainty and with limited information, and they invest in research and collaborations within the hydropower industry to reduce technical uncertainty and to aggregate information for improved decision-making. DOE has core engineering and technology research capabilities and products at its national laboratories, developed to support multiple energy and industrial sectors, that can enhance Corps and Reclamation efforts to improve their knowledge base, condition monitoring, and testing capabilities to support asset management decisions.

- **Hydropower Sustainability**: For the purposes of this MOU Action Plan, sustainability includes, but is not limited to, the management and reduction of the physical, environmental, and ecological footprints of hydropower development and operation—in aggregate (cumulatively) and for individual facilities. It also includes the adequacy of existing and future water availability for hydropower operations. Corps and Reclamation have made major investments over decades to research and characterize environmental impacts, develop and deploy environmental mitigation technologies, and develop management strategies that contribute to the sustainability of their assets and operations, even as science gaps and technological challenges in this area remain. DOE's national laboratories have been engaged in much of that effort—in most cases as providers of fundamental and applied ecological, environmental, and hydrologic science upon which public policy, legal proceedings, and environmental technology solutions for federal and non-federal hydropower assets are based. DOE has also made R&D investments at its national laboratories and other institutions to execute high-risk high-value research that accelerates the development and deployment of fish-friendly, water quality, and other environmental mitigation technologies.

- **Quantifying Hydropower Capabilities and Value in Power Systems**: DOE has invested significant federal R&D resources to partner with power system planners and stakeholders to clarify, model, and quantify future power system planning and operational challenges that will emerge in future power systems as the portfolio of generation and storage assets continues to change. Existing and future hydropower assets are included in the scope of those efforts, as are the quantification of flexible dispatch and ancillary services capabilities that make those hydropower assets valuable beyond simply providing energy. Corps and Reclamation, and the power marketing administrations (PMAs) must continue to operate reliably and meet their existing customer obligations as they gain clarity about how demands on their assets and power systems will evolve, particularly with respect to increasing penetrations of variable renewable energy technologies within power systems. They are key partners and stakeholders in power systems modeling and planning research to resolve hydropower capability and value.

---

[4] Corps maintains and operates federal hydropower assets in multiple river basins throughout the United States, with power from those assets marketed and delivered by the four power marketing administrations. Reclamation maintains and operates federal hydropower assets in river basins across the western states, with power marketed and delivered by WAPA and BPA. The Tennessee Valley Authority, which is not a party to the MOU, maintains and operates federal hydropower assets in the Tennessee River Basin and markets and delivers the power from those assets directly to its customers.

- **Information Sharing, Coordination, and Strategic Planning**: The MOU agencies will continue to share asset configuration data, operational data, and environmental impacts data, within the constraints of infrastructure security and power marketing business sensitivity, to enable regional- and national-scale analyses and research relevant to the entire U.S. fleet of hydropower assets. The agencies will also undertake more intensive efforts to catalog and share past and ongoing internal research efforts among their staff and to disseminate research outcomes and products to the broader U.S. hydropower industry. The agencies will continue to coordinate their activities within the Federal Inland Hydropower Working Group (FIHWG), and will initiate a joint effort to improve the efficiency of permitting for hydropower development. Finally, DOE has undertaken the collaborative development of a long-term national Hydropower Vision for the benefit of the broad U.S. hydropower community. Corps and Reclamation technical staff and managers are key participants in the technical task forces and review committees for that effort.

The sections that follow describe, within each of these areas, the specific action items on which the MOU agencies plan to collaborate to accomplish their goals.[5] These activities are focused on addressing major issues facing both the federal hydropower sector and the larger hydropower industry today, and helping to promote further development of sustainable hydropower resources in the future.

---

[5] This action plan is not exhaustive, as other activities may be considered and initiated in the future. Nor does this document represent a definitive commitment or obligations of funds by any of the agencies.

ACTION ITEMS

Technology Development

CONTINUING: Advanced Hydropower Technology Development Research

Needs Statement: Hydropower technologies have been developed and utilized for many decades, but there is still considerable opportunity for the development and deployment of innovative technologies that produce power more efficiently, significantly reduce costs, increase sustainability, and/or enable or increase development at sites where energy generation has previously been unfeasible or severely limited. To this end, the three agencies will work together where possible within their respective missions to facilitate the use of federal infrastructure (owned or operated by Corps or Reclamation) to conduct testing and demonstrations of innovative, advanced technologies (developed by industry in partnership with DOE).

Description: In April 2011, DOE and DOI announced a joint funding opportunity announcement (FOA)—the Advanced Hydropower Technology Development FOA—which offered financial assistance to projects that develop and demonstrate innovative hydropower technologies that can produce power more efficiently, reduce costs, and increase sustainable hydropower generation at sites not previously considered practical. Sixteen projects totaling nearly $17 million were selected to receive an award. Of those awards, three demonstrations are being conducted at Reclamation sites, two of which were jointly funded by DOE and Reclamation. The two jointly funded demonstrations are the SLH-100 Demonstration Project at Monroe Drop (Natel Energy, Inc., under DOE Grant DE-EE0005420) and Harnessing the Hydro-Electric Potential of Engineered Drops (Percheron Power, LLC, under DOE Grant DE-EE0005428), and the third project at a Reclamation site (funded solely by DOE) is the 45 Mile Hydroelectric Project (Earth by Design Inc., DOE Grant DE-EE0005430).

Timeline/Level of Effort: Two out of the three demonstration projects are expected to complete their installations in fiscal year (FY) 2015, with the third to be completed in FY 2016. As R&D is conducted by DOE in the future, further opportunities for collaborative demonstrations may arise.

Scope/Location: Both the Demonstration Project at Monroe Drop and the 45 Mile Hydroelectric Project are located on the North Unit Irrigation District's Main Canal near Madras, Oregon, while the demonstration site for the Harnessing the Hydro-Electric Potential of Engineered Drops project is located in Montrose, Colorado, at the South Canal.

Final Deliverable: Deliverables include performance testing results to demonstrate energy cost reductions that could be replicated at other sites.

Lead Agency and Participating Agencies/Organizations: This effort is led by DOE with involvement from Reclamation.

Participants are as follows:

DOE: Wind and Water Power Technologies Office (WWPTO)
Reclamation: Power Resources Office, Pacific Northwest Regional Office, Upper Colorado Regional Office
Corps: N/A

NEW: Development of Best Practices Guidance Manuals for Canal Testing of Hydrokinetic Technologies

Needs Statement: Irrigation canals in the western U.S. have promise as sites for the deployment and testing of both low-head hydropower and various hydrokinetic (HK) technologies. With regard to HK technologies, generation of electricity from the flowing water in canals without building conventional hydropower (CH) infrastructure (i.e., dams or head tanks) shows potential to support local electricity needs with minimal regulatory or capital investment. However, very little is known about how the installation of multiple HK devices in canals will affect normal operations and water delivery, especially if there are also existing hydropower plants that operate as part of the system.

Description: The objective of this work is to support HK turbine performance tests in federal canals and determine the impacts of HK devices on water operations through both field measurements and numerical modeling. One major goal of this work is to establish test protocols and best practices for conducting tests/studies that can guide development of the canal-based HK industry. This work will also develop the information needed to predict baseline cost-of-energy estimates for HK electricity generation and will enable future studies to assess the cost/benefit between pursuing HK or more CH projects in locations where both may be suitable.

Timeline/Level of Effort: This effort began late in quarter four of FY 2013 and is anticipated to be complete in FY 2015.

Scope/Location: This work is focused on two primary areas of study:

1. HK device performance testing and analysis with the purpose of characterizing inflow conditions, power performance, and flow recovery in device wake.

2. HK device/array effects on water operation with the intent of characterizing water system energy losses and alterations to the water surface profile (i.e., backwater and drawdown effects).

At present, the Roza Canal in Yakima, Washington (Yakima River Basin) is the primary testing site; however, this could be expanded as other water districts express interest in local site studies. The results of this work will be transferable to any region and therefore applicable nationally.

Final Deliverable: This project will deliver best practices guidance manuals that establish a set of test protocols and recommended methodologies for conducting tests/studies that can be used in considering the development of the canal-based HK industry.

Lead Agency and Participating Agencies/ Organizations: This is a collaborative project being led by Reclamation (testing activities at the Roza Canal) with participation from Corps and DOE. DOE will lead development of the best practices manuals for canal-based HK tests/studies through Sandia National Laboratories (SNL).

Participants are as follows:

DOE: WWPTO, SNL
Reclamation: Science and Technology Office, Power Resources Office, Pacific Northwest Regional Office, Hydraulic Investigations and Laboratory Services Office
Corps: Hydroelectric Design Center (HDC)

NEW: Evaluating and Testing of Superconductor Technology for Hydroelectric Generators

Needs Statement: To date, superconductor technology has not been tested or utilized in hydroelectric generators in the U.S. Preliminary discussions indicate that the use of this technology can potentially reduce losses and increase generator output.

Description: DOE and Reclamation will work together with industry to evaluate a superconductor hydroelectric generator field winding system. This work, if successful, could be followed by the construction and testing of a superconductor hydroelectric generator field winding pole. If testing demonstrates value, a site at a Reclamation facility could be selected and a plan for field installation developed. The ultimate goal is to determine if the application of superconductor technology to hydroelectric generators is a viable and economical means of increasing renewable energy capacity, generation, and reliability.

Timeline/Level of Effort: This project would run from FY 2015 to FY 2017.

Scope/Location: The project's initial scope would include evaluation of such systems, which, if successful, could lead to modeling and testing at a Reclamation hydropower plant in the western United States. If successful, the use of superconducting generators in hydropower facilities could have worldwide application.

Final Deliverable: The initial-stage deliverables will be analyses and reports, potentially followed by lab tests of a superconductor field pole and reports on performance testing. If opportunity exists and the determination is made to install the system in an existing generator, deliverables will include a hydroelectric generator field winding system, as well as test plans and reports describing how that system performed.

Lead Agency and Participating Agencies/Organizations: Activities will be jointly led by DOE and Reclamation with participation from Corps.

Participants are as follows:

DOE: WWPTO, Advanced Research Projects Agency-Energy, Bonneville Power Administration (BPA)
Reclamation: Power Resources Office, Pacific Northwest Regional Office, Science and Technology Office
Corps: HDC, Engineer Research and Development Center

Asset Management

CONTINUING: Hydropower Modernization Initiative

Needs Statement: Within Corps' hydropower fleet, many of the components are approaching the end of their original design or service life and, without major investments in renewals and replacements, are subject to increasing risk of failures. The Hydropower Modernization Initiative (HMI) utilizes asset management principles to support decisions on where to make modernization investments that provide the greatest return on investment, control the overall exposure to the risks of major generating equipment failure, and support the creation of a coordinated, organization-wide, long-term strategy for maintaining the reliability, efficiency, and safety of their hydropower fleet.

Description: Corps is currently in the implementation phase of its HMI effort, which seeks to prioritize modernization activities at its facilities. A total of 54 plants are included in the initiative across three PMA regions. Agreements have been finalized in two regions to fund the recapitalization of HMI projects. These projects will result in efficiency improvements and/or capacity upgrades at most of the 54 plants without any increases in streamflow. Efficiency and capacity gains will be quantified on a project-by-project basis. There is a tremendous amount of interest from the private sector to provide capital for improvements of Corps hydropower facilities. Efforts to acquire the authority for Corps to accept private investments are ongoing.

Timeline/Level of Effort: The HMI Asset Investment Planning tool has been developed. It is being utilized to develop an annual HMI Implementation Strategy that informs Corps hydropower budget development decisions. In July 2013, Corps also completed *The Hydropower Resource Assessment at Non-Powered USACE Sites*, which took a much more rigorous look at the hydropower potential of non-powered Corps facilities than a previous study completed in May 2007.

Scope/Location: The HMI data will cover Corps and Reclamation facilities across the nation.

Final Deliverable: The HMI will yield an updated report on potential hydropower development at existing federal facilities and a list of Corps and Reclamation facilities and sites that are best suited for upgrades or projects to increase generation in a sustainable manner.

Lead Agency and Participating Agencies/Organizations: Corps will lead the HMI effort with support from Reclamation, as well as DOE and its national laboratories.

Participants are as follows:

    DOE: WWPTO, ORNL
    Reclamation: Power Resources Office
    Corps: HDC, HDC's Hydropower Analysis Center (HAC)

NEW: Hydropower Asset Management Research

Needs Statement: Hydropower asset management is central to the mission of federal hydropower operators. However, data, knowledge, and technological gaps exist within hydropower asset management best practices, which can lead to lower efficiencies, higher operations and maintenance costs, more frequent unit and plant outages, and ultimately, reduced generation.

Description: The MOU agencies will focus a portion of their respective research, capital, and operations and maintenance (O&M) resources on these gaps. DOE research efforts in this area will leverage current asset management and analysis activities that are ongoing within the engineering and operations staff of Reclamation, Corps, PMAs, and interested non-federal hydropower operators. DOE's national laboratories and their academic partners will contribute expertise gained from cost and technology modeling in hydropower, wind energy, solar energy, vehicle technology, defense and national security, and other industrial domains. Key industry groups will be engaged to contribute advice and guidance, including the National Hydropower Association (NHA) Operational Excellence Task Force, the NHA Hydraulic Power Committee, and the Hydroelectric Productivity Committee of the Electric Utility Cost Group. These resources would be combined in a multi-year effort to collect enhanced condition-monitoring data, reliability data, cost data, and dispatch data at selected demonstration facilities.

The near-term effort in Asset Management Research activities under this MOU will be focused in two areas:

Flexible Dispatch Impacts Analysis: The goal is to assess reliability and O&M cost implications of "flexible dispatch," defined here as increased hydroelectric machine cycling to provide greater amounts of ancillary services to power systems and allow increasing penetration of variable renewables. Multiple research questions are associated with this issue, with the two fundamental questions being: (a) within current hydropower O&M information, is it possible to detect an increase in O&M costs or degradation of reliability for a facility or a hydroelectric unit and (b) does increased machine cycling cause an increase in O&M costs or degrade reliability for a facility or a hydroelectric unit? Beyond flexible dispatch, other potential causes of reliability and beyond-capital-cost impacts include deployment of new technology, new decision support systems, or new operating paradigms. The investigative framework to be developed within this effort will be applicable to the detection of these causal relationships as well. More generally, the proposed interagency and industry collaboration will develop a monitoring framework and analysis tools for characterizing the reliability and cost impacts of significant changes in hydropower dispatch, deployment of advanced technology, and process improvements. In this framework, the focus is on how the reliability and costs of individual hydropower units and facilities are affected by these factors, rather than how these assets contribute to power system reliability (for a discussion of flexibility and reliability provided to the bulk power system, see the section that follows on *Quantification of Ancillary Services Available from Hydropower Assets*).

Flow Measurement Availability, Accuracy, and Value: The goal is to evaluate the technical feasibility, benefits and costs of developing and installing advanced flow measurement systems to improve the efficiency of hydropower facilities. The first asset management question is how much investment is economically beneficial for an owner to make in upgrading flow measurement capabilities for the individual hydroelectric units within a facility. Within this research scope, DOE will begin to develop—with Corps and Reclamation collaboration—computational tools for modeling the accuracy and cost of current and future flow measurement sensors systems for hydroelectric units. Achieving long-term water-use efficiency and sustainable water management objectives requires active monitoring and control of hydropower unit operations. The specification and availability of accurate flow rate measurement technology is a primary factor in monitoring and controlling the instantaneous efficiency of hydropower energy production amidst multiple constraints on hydropower asset operations. Reducing the uncertainty of unit-specific flow rate measurement and performance will enable U.S. hydropower assets to produce more energy from available water. It will also provide improved information and direct benefit to other Hydropower MOU activities, including (1) *Quantification of Ancillary Services Available from Hydropower Assets*, (2) the *Water Use Optimization Toolset for Hydropower System Scheduling*, and (3) the aforementioned *Flexible Dispatch Impacts Analysis*.

Timeline/Level of Effort: Flexible Dispatch Impacts Analyses and Flow Measurement Research are envisioned to occur over three to five years, and may include longer periods of monitoring for long-developing or rare-event impacts of flexible dispatch. Activities in FY 2013–2014 included scoping and preliminary modeling that will inform a collaborative, integrated effort. Efforts in FY 2015–2016 will focus on retrospective case study analyses of configuration and O&M data to determine the benefits and costs of flow measurement technology and the impacts of flexible dispatch. Within the flow measurement effort, further development of reference geometries, turbulent flow field modeling, and sensor testing will occur.

Scope/Location: Sites for retrospective studies and pilot testing would be selected from among Corps, Reclamation, or non-federal assets. The outcomes of these initial studies will be applicable to hydropower facilities nationwide. Preliminary FY 2014–2015 R&D activity for the Flexible Dispatch and Flow Measurement activities will be concentrated at DOE's national laboratories (Oak Ridge National Laboratory [ORNL] and Pacific Northwest National Laboratory [PNNL]); Corps' HDC; and Reclamation's Technical Service Center (TSC).

Final Deliverable: Important long-lived deliverables of the Flow Measurement effort will be (1) a peer-reviewed reference flow field specification and dataset that will accelerate testing and deployment and lower costs for HFM sensor developers, and (2) a publicly-available flow measurement value analysis tool that will accelerate deployment of advanced flow measurement technology throughout the U.S. hydropower fleet. The Flexible Dispatch Impacts activity will yield peer-reviewed, multi-agency *Hydropower Asset Reliability and Cost Impacts Detection Framework Report* and a peer-reviewed, multi-agency *Flexibility-Reliability-Cost Statistical Findings Report*.

Lead Agency and Participating Agencies/Organizations: DOE will lead the Flow Measurement and Flexible Dispatch Impacts activities with participation from Corps, Reclamation, and input from the PMAs.

Participants are as follows:

DOE: WWPTO, ORNL, PNNL
Reclamation: Power Resources Office, Science and Technology Offices, Hydropower Technical Services
Corps: HDC, HDC's HAC

NEW: Hydropower Operational Data Analysis and Archiving Guidelines

Needs Statement: Analyses of hydropower operations (including those intended to address value, operational performance, policy impacts, market participation, environmental impacts, and other trending) are often limited in scope, detail, and actionable findings because the operational data to support the analyses are less than complete, of insufficient detail, non-existent, or not readily available in a standardized format. Many Hydropower MOU activities described herein will rely on operational data for important analytical findings and may create data that should be maintained in correlation with historical operational data. These activities include the *Quantification of Ancillary Services Available from Hydropower Assets, Water Use Optimization Toolset for Hydropower System Scheduling, Hydropower Asset Management Research, Water Quality Scheduling for Hydropower Releases (Cumberland and Columbia Basins), National Hydropower Asset Assessment Program, Hydropower Modernization Initiative,* and *Assessment of Climate Change Impacts on Water Resources and Federal Hydropower.* It is reasonable to assume that opportunities exist for improving the efficiency of data collection, quality assurance, and exchange among these efforts by developing standards for archiving and documentation.

Description: DOE, Reclamation, and Corps will seek to establish a core set of analyses and accompanying data specifications (granularity, plant/unit breakouts, time step, metadata requirements, and other attributes) that are useful and applicable to all U.S. hydropower facilities and generating units. Other hydropower-related agencies (e.g., PMAs and Tennessee Valley Authority [TVA]) will be included in a needs assessment for core hydropower analyses capability. This needs assessment will be disseminated to non-federal hydropower operators for comment. The set of core analyses will be translated into a detailed specification of a set of operational and plant configuration metadata that describe the operational data and plant/unit parameters that are required to support the core analyses. The specification may include a data format (e.g., an XML-derived specification) for core data that would facilitate exchange of data and automated analyses. Participating agencies and institutions may use the specification to establish condition monitoring; scheduling; and other data collection, assimilation, and archiving processes that meet their management needs. Use of the specification by participating agencies and institutions may also enable aggregation of datasets and analyses across the federal and non-federal hydropower fleets to provide more comprehensive findings from fleet-wide analyses.

Timeline/Level of Effort: The drafting and iterative review and revision by participating institutions will take approximately one year to complete. DOE and its national laboratories will facilitate the drafting process. Reclamation and Corps will provide staff time to review and comment on the draft documents.

Scope/Location: The staff effort for this collaboration will come from ORNL, Reclamation's TSC, and Corps' HDC.

Final Deliverable: A formal technical report will be prepared to document the core analyses set and associated hydropower data specification. Pending further discussion among MOU partners, one or more case studies and example datasets may be developed and made publicly available.

Lead Agency and Participating Agencies/Organizations: DOE will facilitate the drafting process. Reclamation and Corps staff will provide review and revision of the documents.

Participants are as follows:

```
DOE:          WWPTO, ORNL
Reclamation:  Power Resources Office
Corps:        HDC's HAC
```

NEW: Sharing of Best Practices and Performance Metrics to Achieve Operational Excellence

Needs Statement: Hydropower owners and operators conduct business with priority objectives of safety and health; environmental impact minimization; and reliable, cost-competitive production. They manage equipment, staff, and processes to achieve these outcomes, but may do so without the benefit of experience and best practices developed by peer facilities or institutions. They contend with aging equipment and scarce capital improvement resources, and face evolving water and power system demands on facilities and operations staff that engender altered operational schemes and unforeseen challenges. Without recognition and attention to mitigation, these challenges can decrease safety, increase impacts, and decrease the value that hydropower assets provide. Industry forums do exist to promote the sharing of best practices, lessons learned, and performance metrics, but they are neither comprehensive nor systematic in their coverage of equipment (technology), staffing, and process issues. They are also less than comprehensive in their coverage of the entire fleet of U.S. hydropower facilities. Thus, there are opportunities to improve the operations and increase the value of federal and non-federal hydropower assets nationwide through more effective and comprehensive information sharing.

Description: The MOU agencies will work together to create or strengthen systematic forums for sharing hydropower operating experiences, significant events, best practices, lessons learned, and performance metrics among the hydropower industry. This activity will increase the hydropower industry's standard of performance by enabling valuable insights and relevant information gained at a facility to be efficiently reported, validated, classified, cataloged and made available to staff at other facilities. The inclusion of this initiative in the MOU Action Plan does not obligate the MOU agencies to begin sharing information immediately. As they work toward a comprehensive information-sharing objective, the MOU agencies will need to confront issues of business sensitive or critical infrastructure data sharing, potential reticence to report failures and lessons learned, and disparate or conflicting terms of reference (bylaws) for existing industry forums in which the MOU agencies participate. The intent of this MOU activity is not to supplant existing industry forums, but to establish a goal of coordination and expansion of forum activities to achieve comprehensive information sharing and peer performance comparison across the entire fleet of U.S. hydropower assets.

Timeline/Level of Effort: This activity will be ongoing, with an initial effort in the first six months to characterize the information-sharing forums in which MOU agencies participate and identify opportunities for coordination and expansion of coverage.

Final Deliverable: DOE and ORNL will prepare a brief technical report on operational excellence information-sharing and use across the U.S. federal and non-federal hydropower fleets, with specific attention to trends in the amount of information shared and used and the comprehensiveness of cataloged information.

Lead Agency and Participating Agencies/Organizations: DOE will take the lead in characterizing existing information-sharing forums, assessing the state of information sharing, and identifying opportunities for enhancement. DOE and its national laboratories will also provide data analytics capabilities when needed to enhance the usefulness of information-sharing forum databases. Reclamation and Corps will provide synopses of their information-sharing efforts to DOE and provide concurrence review of DOE analyses and reports on the subject of hydropower information to achieve operational excellence.

Participants are as follows:

DOE: WWPTO, ORNL
Reclamation: Power Resources Office
Corps: Corps Headquarters Operations

Hydropower Sustainability

CONTINUING: Assessment of Climate Change Impacts on Water Resources and Federal Hydropower

Needs Statement: The SECURE Water Act of 2009 directed Reclamation and DOE to produce assessments of the effects of climate change on water resources (Section 9503) and hydropower (Section 9505). The initial assessments were completed in 2011, and they are to be repeated again in 2016. Planning for the second round of these assessments began in FY 2014. Interagency coordination and collaboration will be an important part of planning these assessments.

Description:

9503: The next 9503 assessment of water resources will update climate change impacts to water supply, using new climate projection information made available since 2011. It will also leverage vulnerability assessments conducted through the West-Wide Climate Risk Assessment (WWCRA) Impact Assessments and Basin Studies to identify risks to various water demands and water management operations in major Reclamation river basins. In addition, the next 9503 report will discuss adaptation strategies that have been developed to mitigate for the impact of changes to water supply due to climate change. The updates to water-supply impact will include changes in atmospheric conditions, such as temperature and precipitation, as well as changes in hydrologic conditions that can affect streamflow timing, quantity of runoff, snowpack conditions, and impacts to groundwater supplies. Reclamation will also provide a more detailed assessment of impacts to crop water demands resulting from warmer temperatures and changes in growing seasons. The WWCRA Impact Assessments and Basin Studies will provide the basis for characterizing impacts to water management operations that may affect water deliveries, hydroelectric power generation, ecological resources, recreation, and flood control management.

9505: The second 9505 assessment of hydropower impacts will include an updated description of the federal power system in the United States and of the power marketing practices of the PMAs. Future climate impacts to the federal power system will be estimated using a series of climate models, including General Circulation Models of the globe, regional downscaling models with increased spatial resolution, and hydrologic models that represent energy and water dynamics at a spatial scale relevant to federal hydropower projects. New and existing data in the ORNL National Hydropower Asset Assessment Program (NHAAP) database will be used to parameterize the assessment models. Assessment endpoints (primary output variables) will be designed collaboratively with the federal agencies to ensure applicability to the federal hydropower systems. Relative to the first 9505 assessment, this new work will endeavor to examine shorter time-intervals (e.g., sub-annual generation patterns); frequency and magnitude of extreme hydrologic events; interactions between climate effects and competing water uses that may affect future water availability; and indirect effects, such as water temperature, groundwater interactions, evaporative losses, changing environmental protection requirements, and climate-induced demand for electricity.

Timeline/Level of Effort:

9503: In FY 2014, Reclamation updated the water-supply analysis using new climate projections and developed a structured approach for aggregating the previous WWCRA Impact Assessments, Basin Studies, and other climate change studies conducted throughout the western 17 states into a single, comprehensive, and cohesive report that meets the needs of 9503. A draft 9503 assessment report will be completed toward the end of FY 2015, allowing the internal and peer-review processes to take place in FY 2016.

9505: Efforts in FY 2014 focused on research and design of the new assessment methods. Efforts in FY 2015 will focus more on application of new methods, analysis, and documentation of results. A draft assessment report will be completed toward the end of FY 2015, which will then be subjected to rigorous peer review.

Scope/Location: The geographic scope of this work will be nationwide, covering all river basins containing federal hydropower systems that produce electricity marketed by the federal PMAs. This includes 132 hydropower projects owned and operated by either Corps or Reclamation.

Final Deliverable: This work will produce multiple end products, including technical assessment reports and Reports to Congress that are responsive to Sections 9503 and 9505 of the SECURE Water Act.

Lead Agency and Participating Agencies/Organizations: Reclamation is responsible for Section 9503 products on other aspects of water resources in the western states. DOE is responsible for Section 9505 products on hydropower effects. Corps will also provide interagency collaboration associated with climate impacts at their projects. Other federal and state agencies will be consulted in the design and implementation of the assessments, including the National Oceanic and Atmospheric Administration (NOAA), U.S. Geological Survey (USGS), and members of the FIHWG that was set up under this MOU. Participants are as follows:

DOE:	WWPTO, ORNL, BPA, WAPA, SEPA, SWPA
Reclamation:	Program Management Office, Science and Technology Office
Corps:	Institute for Water Resources

CONTINUING: Basin Scale Hydropower and Environmental Opportunity Assessment Initiative

Needs Statement: The Basin Scale Opportunity Assessment (BSOA) initiative originated as an action item in the federal Hydropower MOU. The purpose of the MOU is to "help meet the nation's needs for reliable, affordable, and environmentally sustainable hydropower." The MOU agencies, while recognizing that hydropower is the largest source of renewable electricity in the nation, emphasized that efforts to increase hydropower generation must also aim to improve environmental conditions in our nation's rivers and watersheds. Accordingly, the goal of the BSOA initiative is to develop and implement an integrative approach for the assessment of hydropower and environmental opportunities at a basin scale.

Description: The goal of this initiative is to develop an approach to hydropower and environmental assessment that emphasizes sustainable energy systems within the context of basin-wide environmental protection and restoration, focusing on low-impact or small hydropower and related renewable energy. Assessments will treat the basin as an integrated system and identify opportunities or scenarios in which hydropower generation or value could be increased, while simultaneously improving environmental conditions in the basin. A wide variety of interested hydropower stakeholders can utilize assessment products to identify and evaluate opportunities to meet both environmental and hydropower goals within large or complex river basins.

By FY 2015, this project will have completed assessments in at least four basins, developed and documented the assessment approach, and exported the approach to interested stakeholders. The ultimate goal of this initiative is to develop and test the ability to integrate hydropower and environmental goals to guide collaborative decision-making within river basins.

Timeline/Level of Effort: The majority of the remaining work under this project is scoped to occur during FY 2014-2015, with final results being publicized during the beginning of FY 2016.

Scope/Location: This is a national-scale project, with specific activities occurring in four basins: The Deschutes Basin (Oregon), the Roanoke Basin (Virginia and North Carolina), the Connecticut Basin (Connecticut, Massachusetts, New Hampshire, and Vermont), and the Bighorn Basin (Montana and Wyoming).

Final Deliverable: The final deliverables will include a detailed assessment in the Deschutes basin and scoping-level assessments in the Roanoke, Connecticut, and Bighorn basins; documentation of the opportunity assessment approach; and Web-based tools for data visualization and display.

Lead Agency and Participating Agencies/Organizations: DOE is leading this effort through PNNL, with Corps, Reclamation, and other national stakeholders serving on the project's steering committee. Participants are as follows:

DOE: WWPTO, PNNL, ORNL
Reclamation: Power Resources Office, Pacific Northwest Regional Office, Great Plains Regional Office
Corps: Wilmington District, New England District

CONTINUING: Validation and Analysis of Alden Fish-Friendly Turbine

Needs Statement: Though successful scale model tests for efficiency and environmental performance have been carried out on the newly developed Alden fish-friendly turbine, prototype demonstration and testing at an operational hydropower facility has not yet occurred. Without a full-scale operational demonstration, the performance and facility-wide costs and benefits of the turbine cannot be accurately validated.

Description: The MOU agencies intend to continue exploring various options to test the Alden fish-friendly turbine. This will largely depend on the development of external partnership opportunities and the availability of an adequate facility and resources. In the near term, it will be beneficial for the MOU agencies to publicly document what efforts have occurred to date in an attempt to secure a demonstration site for the turbine, as well as share information and lessons learned about what difficulties have been encountered thus far. This will help to better educate and inform potential partners.

Timeline/Level of Effort: While efforts to secure a site and conduct demonstration tests of the Alden turbine will continue to occur, assembly of a report on lessons learned to date will also occur during FY 2015, with the release of the report at the end of FY 2015.

Scope/Location: The search for an appropriate demonstration site will be nationwide. The lessons learned report will encompass insights from the project's inception.

Final Deliverable: A demonstration of the Alden fish-friendly turbine at an appropriate site and results from that demonstration are still possible, but in the near term, a report on lessons learned to date will be delivered in FY 2015.

Lead Agency and Participating Agencies/Organizations: DOE will be leading these efforts with input from Corps and Reclamation. Participants are as follows:

DOE:	WWPTO, BPA
Reclamation:	Power Resources Office, Science and Technology Office
Corps:	HDC

NEW: Advanced Design Tools and Criteria to Improve Hydro-Turbine Biological Performance

Needs Statement: Advanced design tools are needed that can reliably predict—prior to manufacture and installation—biological performance of hydro-turbines and other hydropower plant system components. Development and delivery of design tools will have national impact by reducing design time, regulatory risks, cost, and most importantly, can significantly improve the environmental performance of both new hydropower turbines and upgrades/replacements of turbines at existing facilities. These tools will lead to new designs that maintain or increase generation while improving the ease of meeting regulatory compliance requirements (e.g., fish passage, water quality). Furthermore, advanced design tools will reduce the need for additional live fish or sensor fish testing and reduce the significant costs associated with field data collection.

Description: The project goal is to develop advanced hydro-turbine design tools, based on computational fluid dynamics models and traits-based assessment of species of concern to predict the biological performance in the design phase for new hydro-turbines, or refurbishment of existing hydro-turbines nationwide where fish passage is a regulatory issue.

The project will start by building on the initial version of the biological performance assessment (BioPA) tools developed by PNNL during 2010-2013, extending the tools to other types of turbines (e.g., Francis, Bulb), and confirming model performance through comparisons to existing live fish and sensor fish data. Working with MOU partners, DOE will also begin to assemble turbine data for a larger number of plants and specifically small hydro-turbines. Working groups with industry, operators, and regulators will be formed to establish priorities, exchange information, and initiate technology transfer.

Data gaps in biological design criteria were identified during the initial project phase in FY 2014. Species of concern will be identified in order to develop refinements in the dose-response relationships incorporated in the design tools. Closing data gaps and developing refined criteria may require additional laboratory biological testing conducted in coordination with PNNL, ORNL and other partners. Improved hydraulic and fish simulation will be included in the tools to account for key features, such as moving blades, turbulence, and fish-particle mass. The project will also address issues specific to small hydro-turbines and associated installations.

In the final project year, refinements to the biological design criteria and comparisons with newly available live fish and sensor fish field data for large and small hydro-turbines will be completed. Improved hydraulic and fish simulation capabilities will be incorporated into a second-generation version of the BioPA tools. These tools will be made available to industry and agencies through technology transfer workshops

Timeline/Level of Effort: The project is scoped to occur from FY 2014 to 2017.

Scope/Location: The technology developed from this activity will be applicable at hydropower facilities nationwide. Demonstration sites would also be selected from among Corps, Reclamation, and non-federal (e.g., Wanapum Dam) hydropower projects in the Pacific Northwest region and then expanded to include other facilities nationwide. Initial R&D activity will be concentrated at DOE's national laboratories (PNNL and ORNL) and at Corps' HDC.

Final Deliverable: The end product will be an advanced set of hydro-turbine design tools created in coordination with turbine developers, allowing them to include biological performance in their standard design process and in communication and coordination with regulatory agencies.

Lead Agency and Participating Agencies/Organizations: DOE will lead this effort with participation from Reclamation and Corps. Participants are as follows:

DOE:	WWPTO, BPA, PNNL, ORNL
Reclamation:	Science and Technology Office
Corps:	HDC

NEW: Water Quality Scheduling for Hydropower Releases (Cumberland and Columbia Basins)

Needs Statement: Hydropower schedulers and operators typically rely on experience and conservative guidelines in scheduling flow releases to ensure compliance with downstream temperature and dissolved gas constraints. Guidelines are increasingly based on findings and insights from high-fidelity hydrodynamic and water quality models, but there are presently no mechanisms to directly include the spatiotemporal fidelity and dynamics of these models in flow release scheduling. These models are based on highly nonlinear mathematical equations that describe the dynamics of river flows and water quality, but such equations are too complex to integrate directly into real-time hourly scheduling systems for hydropower assets. This research will provide a set of simplified equations, derived from the complex equations and calibrated from high-fidelity water quality model outputs that can be integrated into hydropower operational scheduling tools such as the RiverWare system or the real-time hourly scheduling component of the Water Use Optimization Toolset (WUOT). The success and research challenge of this "model-reduction" strategy requires a balance between adequately capturing the variability of water quality in waterways (so as to comply with water quality targets) and simplifying the mathematical formulation of water quality dynamics enough to allow for real-time accurate solutions of the hourly scheduling problem.

Description:

Cumberland Basin: The initial development and testing of this approach to water quality scheduling is occurring in the Cumberland River for two hydropower facilities—Cordell Hull and Old Hickory—operated by the Nashville District of Corps. Vanderbilt University is a research partner in this effort. The focus in the Cumberland River effort is on leveraging the spatiotemporal fidelity of calibrated CE-QUAL-W2 models of reservoir and tailwater temperature and dissolved oxygen into simplified dynamics for real-time scheduling for the Cumberland River hydropower system. The Cumberland River provides an archetypical context of a highly regulated, water quality constrained, daily power-peaking, multi-purpose reservoir system in which to develop and demonstrate techniques for water quality scheduling and co-optimization of energy and water quality objectives.

Columbia Basin: Similar work is also occurring within collaboration between ORNL, Reclamation, and the Corp's Northwest Division staff to address dissolved gas scheduling constraints for multiple facilities within the Federal Columbia River Power System (FCRPS) and non-federal projects on the Mid-Columbia River. The University of Iowa and the University of Colorado-Boulder are research partners in this work. These collaborators are developing simplified methods to determine and manage localized and system-wide total dissolved gas (TDG) concentrations in Columbia River Basin reservoirs and tailwaters. The Columbia Basin experiences significantly more spill than some basins amidst sometimes conflicting drivers for fish passage and TDG management. A system-wide, real-time, data-driven approach is being developed to model spill maximums for fish survival at dams, requiring hourly meteorological information and hourly flow, spill quantity, water temperature, and TDG data received via satellite from fixed monitoring sites in the Columbia River.

Timeline/Level of Effort: TDG scheduling research for the Columbia Basin began in 2012 and will likely conclude with products delivered in 2015. Water temperature and dissolved oxygen scheduling research for the Cumberland Basin began in 2013 and is scheduled to deliver products in 2016.

Scope/Location: The methodologies being developed and demonstrated will be applicable to river regulation throughout the U.S. and the world wherever water quality concerns exist. For the demonstration areas, the Cumberland River begins in Eastern Kentucky and traverses middle Tennessee before emptying into the Ohio River near Paducah, Kentucky. The Columbia River Basin includes major portions of Alberta, British Columbia, Washington, Oregon, Idaho, and Montana, and minor portions of Nevada, Utah, and Wyoming.

Final Deliverable: The formulation and validation of the methodologies demonstrated in the Cumberland and Columbia Basins will be published in peer-reviewed technical papers in one or more water management journals. The process, challenges, and outcomes of deploying the methodologies in the Corp's Nashville District-FCRPS and Mid-Columbia hydropower operations will be documented in ORNL technical reports. The Columbia Basin partners will also produce a historical database of correlated hourly hydropower operations and TDG monitoring for multiple FCRPS and Mid-Columbia facilities.

Lead Agency and Participating Agencies/Organizations: DOE will be leading this effort with support from ORNL and inputs from Reclamation and Corps.

Participants are as follows:

DOE: WWPTO, ORNL
Reclamation: Power Resources Office/Environmental Applications & Research Office,
             Science and Technology Office
Corps: Nashville District, Northwestern Division

Quantifying Hydropower Capabilities and Value in Power Systems

CONTINUING: Pumped-Storage Screening Study

Needs Statement: Reclamation is working to increase hydropower capacity and identify options to integrate other forms of renewable energy, such as wind and solar generation, into its energy portfolio. Pumped-storage facilities have the potential to contribute both increased capacity and options for other renewable energy integration. Reclamation recognizes that there may be potential to develop additional pumped-storage projects within the existing federal footprint, provided there is both technical and economic feasibility. Some investigation has been done looking into the potential of developing pumped-storage projects on Reclamation sites; however, none have moved past the assessment study phase, and there are no active projects at this time.

Description: Reclamation will investigate the feasibility of utilizing its existing reservoirs for potential pumped-storage development without negatively impacting existing project operations. Pumped-storage is recognized as one of the most useful methods for regulating variable resources like wind and solar. Because the cost of developing these facilities is considerable, Reclamation investigated the potential for converting existing conventional hydroelectric facilities with an existing forebay and afterbay reservoir to pumped-storage facilities in FY 2013. This study also investigated the potential for adding an additional upper reservoir and new pumped-storage plant to four sites. None of these conversions showed promise. In FY 2014, Reclamation will utilize some of the lessons learned from this study and investigate the potential for adding a pumped-storage configuration to 60 of its reservoirs that have an existing active reservoir storage capacity greater than 100,000 acre-feet. This study will use a systematic approach that will rank the potential for developing pumped-storage at these 60 sites based on a variety of factors, including an operational screening; a facility layout and cost evaluation; transmission and market constraints/opportunities; and environmental, cultural resource, and other institutional constraints.

Timeline/Level of Effort: The project was started in FY 2013, analysis was completed at the end of FY 2014, and Reclamation will be assessing next steps in FY 2015.

Scope/Location: This research will be conducted using Reclamation's reservoirs.

Final Deliverable: The deliverable will be a final report that ranks the potential for pumped-storage at each site. This ranking will allow Reclamation to prioritize additional research for the potential development of pumped-storage at Reclamation assets.

Lead Agency and Participating Agencies/Organizations: Reclamation is leading this effort with support from DOE.

Participants are as follows:

DOE: WWPTO, WAPA, BPA
Reclamation: Power Resources Office
Corps: N/A

CONTINUING: Water Use Optimization Toolset for Hydropower System Scheduling

Needs Statement: Hydropower operators and planners have expressed the need for improved analytical tools to manage the increasingly more complex conflicts between water, power and environmental considerations. There is the potential to significantly improve hydropower system optimization, meaning more energy and grid services from available water, and enhanced environmental benefit from improved planning and coordination of operations at multiple hydropower facilities.

Description: DOE sponsored the development of an integrated set of advanced analytical tools for optimizing water, power, and potentially environmental benefits in hydropower systems. WUOT's potential to improve the performance of hydropower systems (hydraulically linked hydropower facilities arranged in series and parallel within river systems) has been shown at previous demonstration sites. DOE is sponsoring additional testing, demonstration, and technology transfer actions for the WUOT to promote U.S. hydropower operational performance improvement. The development phase of the project was completed in FY 2013. The testing and implementation phase will extend over the next three years. During that time, DOE will sponsor the testing and implementation of the toolset at a diverse set of hydropower systems, requisite modification of the WUOT to meet user needs, and transfer of the toolset to hydropower planners and operators.

Timeline/Level of Effort: During FY 2013–2015, DOE will continue to fund the testing, demonstration, and technology transfer of the WUOT.

Scope/Location: The WUOT will be tested, demonstrated, and transferred at sites across the United States. Demonstrations already completed or planned to date include the following:

- Hydropower system of Seattle City Light with Seattle City Light (new analysis)

- Aspinall Cascade in the Colorado River Storage Project with Reclamation and WAPA (retrospective analysis complete, WUOT application ongoing)

- Oroville Hydropower Facility on the Feather River in California with the California Department of Water Resources Administration (retrospective analysis complete, WUOT application ongoing)

- Conowingo Dam Complex on the Susquehanna River in Maryland with Exelon and PJM Interconnection LLC Administration (retrospective analysis complete, WUOT application ongoing).

Final Deliverable: The final deliverables will include an electronic version of the WUOT, a series of training materials, and documentation of results from the site demonstrations. Additional outcomes include the transfer and use of the WUOT by hydropower operators and planners.

Lead Agency and Participating Agencies/Organizations: DOE is leading the activity with participation from Reclamation and Corps. Additional participants include WAPA, California Department of Water Resources Administration, Exelon Corporation, Seattle City Light, and a technical review team consisting of hydropower operators and planners from across the United States.

Participants are as follows:

DOE: WWPTO, Argonne National Laboratory (ANL), WAPA
Reclamation: Power Resources Office, Science and Technology Office, Upper Colorado Regional Office
Corps: Water Management

NEW: Irrigation Pumping Demand-Response Initiative Guidelines

Needs Statement: Reclamation has designed, built, and constructed pumping plants throughout the western U.S. to move water to areas of demand. These pumping plants are occasionally located at intakes for canals or to transport water over certain topographical features that could not otherwise be overcome. Over Reclamation's service area, Project Use Power delivered for pumping loads accounts for approximately 8% of Reclamation's total hydropower generation. Flexibility and potential for efficiency increases may exist within Reclamation's overall demand and pumping requirements to better manage and integrate renewable power with the electric generation and transmission network. This flexibility could be used to encourage the development of renewable energy, as identified in the *Interior Strategic Plan* and *Reclamation Sustainable Energy Strategy*.

Description: While most of Reclamation's renewable energy efforts have focused on the development of new renewable projects and technologies, additional opportunities may be available through energy conservation. Several utilities, including BPA, have programs in place to improve efficiencies on the demand side of the system. These programs include refurbishment of irrigation pumps and equipment to improve the efficiencies of that equipment. Reclamation will investigate this potential in three phases.

The first phase will prepare an inventory and summary analysis of (1) completed programs to improve the energy use and overall efficiency of pumping plants and irrigation projects, and identify particular irrigation components examined, such as pumps, the methodologies employed for these analyses, and the specific parts of these components, which were replaced and/or refurbished; and (2) identify and describe economic and financial incentive programs that can help facilitate the adoption of more effectual irrigation technologies through increased pumping efficiency, and/or incentive programs that focus on demand management for pumping plants during peak demand periods to provide more grid flexibility.

The second phase will consist of case studies at a selection of Reclamation's irrigation projects to identify where opportunities may exist for efficiency improvements and/or demand management flexibility.

The third phase will utilize the lessons learned in the first two phases to potentially design a program that will promote the implementation of efficiency and demand response activities at Reclamation irrigation projects. Work under this topic will involve collaboration with BPA and WAPA.

Timeline/Level of Effort: The program will begin in 2014 and continue through 2017.

Scope/Location: This research will be conducted using Reclamation's pumping facilities.

Final Deliverable: The end result of this research will be the development of programs that will improve efficiencies and flexibility to Reclamation pumping loads.

Lead Agency and Participating Agencies/Organizations: Reclamation is leading this effort. Reclamation will work closely with BPA and WAPA to gather necessary input that is needed to complete the project

Participants are as follows:

DOE: WWPTO, BPA, WAPA
Reclamation: Power Resources Office, Science and Technology Office, Lower Colorado Region
Corps: N/A

NEW: Quantification of Ancillary Services Available from Hydropower Assets

Needs statement: It is commonly believed that Reclamation and Corps hydropower facilities are capable of supplying large amounts of ancillary services to the bulk power system to support renewable energy development in the United States. While there is some capability to facilitate grid stabilization and higher penetration of variable renewable resources, it seems likely that this ability is more limited than previously assumed, because Reclamation and Corps facilities are subject to numerous operational, environmental, and regulatory constraints. Review of previous large-scale power system studies suggests that the specification and modeling of federal hydropower asset operating priorities and constraints within bulk power system models often do not fully take into account these constraints. As a result, the findings of grid systems modeling and transmission planning studies may be overly optimistic about the flexibility and quantities of ancillary services available from hydropower assets. The focus of this activity is on the capability of hydropower assets to provide ancillary services, rather than the effect that such operations have on individual hydropower assets (for a discussion of the impacts of changes in dispatch practices on individual assets, see the preceding section on *Hydropower Asset Management Research - Flexible Dispatch Impacts Analysis*). DOE has also been directed in the recently enacted Hydropower Regulatory Efficiency Act of 2013 to investigate the amount of ancillary services and flexibility provided by the nation's fleet of pumped-storage facilities, and then deliver a report to

Congress.

Description: The ability of existing federal hydropower and pumped-storage plants to provide ancillary services—such as up regulation, down regulation, pumping (in the case of pumped-storage plants), spinning reserve, and non-spinning reserve—is a function of many interrelated factors, including the following:

1. Physical site, design, and engineering capability of the plant

2. Forebay reservoir elevation and water availability

3. Purpose, allocation, and priority of water release

4. Environmental and other regulatory constraints

5. Regulations and institutions governing the sale of federal power.

The goal is to rigorously quantify the capability of a subset of Reclamation and Corps hydropower and pumped-storage hydropower (PSH) plants to provide ancillary services for a specified range of hydrologic conditions. Given the important role of operational constraints in determining ancillary services provision, these ancillary services provision capabilities will be estimated for (a) the base case [items 1–3 in the list above], (b) environmental and other regulatory constraints [items 1–4 in the list above], and (c) all existing constraints [items 1–5 in the list above].

This work will build upon the current DOE study *Modeling and Analysis of Advanced Pumped-Storage Hydropower in the United States*. The study is led by ANL and the project team has developed detailed dynamic models of advanced PSH technologies (adjustable speed and ternary units) and analyzed technical capabilities of PSH plants to provide various grid services, as well as the value of those services under different market structures and renewable energy penetration levels. The vendor-neutral dynamic models of advanced PSH technologies have been described in five project reports, which are now publicly available. The production cost

---

[5] http://ceeesa.es.anl.gov/projects/psh/psh.html

and revenue analyses were performed for several geographical areas within the western U.S. and are described in the final project report.[5]

Timeline/Level of Effort: The anticipated study period is approximately 18–24 months beginning in FY 2014 (initially focusing on PSH plants to facilitate DOE completing the required Report to Congress).

Scope/Location: Initially, Reclamation and Corps hydropower facilities (could possibly include other federal hydropower, such as TVA).

Final Deliverable: The final deliverables will be methods/tools for ongoing calculation and a final report.

Lead agency and participating agencies: This effort will be jointly led by Reclamation and DOE with participation from Corps.

Participants are as follows:

DOE: WWPTO, ORNL
Reclamation: Power Resources Office, Science and Technology Office
Corps: HDC's HAC

Information Sharing, Coordination, and Strategic Planning

CONTINUING: National Hydropower Asset Assessment Program

Needs Statement: Prior to 2009, a centralized, publicly-accessible source of information on the U.S. hydropower fleet (including both federal and non-federal facilities) did not exist. Opportunities to further develop and improve the country's existing hydropower system (both for increased energy generation and improved environmental performance) could not be readily identified. DOE's NHAAP is needed to identify the current state of the U.S.'s hydropower infrastructure (age, type, ownership, etc.); generation patterns from these assets; and effects of varying hydrologic conditions on generation, as well as to investigate and catalog new hydropower potential.

Description: ORNL maintains the NHAAP as an integrated energy, water, and ecosystem research effort for sustainable hydroelectricity generation and water management. Launched in FY 2010, the NHAAP started with gathering, organizing, and validating the stream network, facility configuration data, historic generation, and water availability data necessary to trend the production and capacity of U.S. hydropower for WWPTO. As resource assessment efforts for non-powered dams and new stream-reach development were executed, layers of environmental and socioeconomic geospatial data were added to the NHAAP to enable researchers to assign environmental and socioeconomic characteristics to existing and potential hydropower sites. An advisory committee for the NHAAP was established and includes Corps and Reclamation representatives as members. The committee holds regular meetings to exchange available data/research from all ongoing efforts.

These data are derived from federally chartered database efforts and include the HydroAmp database maintained by Reclamation/Corps/BPA, the Federal Energy Regulatory Commission (FERC) eLibrary and DamSafety database, Energy Information Administration Forms 860/923 Powerplant and Generation database, the National Inventory of Dams, USGS/Environmental Protection Agency (EPA) National Hydrography Dataset, and USGS National Water Information Service. ORNL has also acquired facility configuration data from individual hydropower operators. In general, environmental and socioeconomic characteristics, aggregate facility capacity, monthly generation totals, and monthly water availability data are publicly available for individual facilities. Facility and unit-level hydropower configuration and performance data within the NHAAP are typically obtained through bilateral exchanges with agency and electric utility staff and treated as privileged or confidential commercial information. Aggregate statistics from such information are available publicly, but individual facility and unit configuration and performance data are not publicly available.

The NHAAP is designed to integrate these data at various scales and serve as a tool for strategic planning and decision-making to assess the current value of the nation's hydroelectric infrastructure, quantify the amounts of energy that could be feasibly extracted, and provide an environmental attribution resource. Both formal meetings and informal communication are taking place between Reclamation, DOE, and Corps regarding these activities.

Using NHAAP data, DOE responded to Section 7 of the Hydropower Regulatory Efficiency Act of 2013 (H.R.267), which calls for a report to Congress on hydropower from conduits—"any tunnel, canal, pipeline, aqueduct, flume, ditch, or similar manmade water conveyance that is operated for the distribution of water for agricultural, municipal, or industrial consumption and not primarily for the generation of electricity." The report was submitted to Congress in 2015.

Timeline/Level of Effort: Ongoing based on need.

Scope/Location: National.

Final Deliverable: An online database: http://nhaap.ornl.gov/.

Lead Agency and Participating Agencies/Organizations: DOE leads this effort with input for the Corps and Reclamation. Participants are as follows:

DOE:	WWPTO, ORNL
Reclamation:	Power Resources Office
Corps:	HDC's HAC

CONTINUING: Federal Inland Hydropower Working Group

Needs statement: The MOU agencies are not the only federal agencies involved in hydropower. To coordinate better information sharing across all of the federal agencies engaged in hydropower, the MOU (Section D) formed the FIHWG.

Description: The FIHWG is made up of 16 federal entities involved in the regulation, management, or development of hydropower resources (including hydrokinetics) in rivers and streams of the United States. The working group convenes quarterly, via staff-level meetings, to update federal agencies on the status of initiatives, efforts, and projects related to hydropower. The meetings are also utilized to update project leads from DOE, Corps, and Reclamation on the status of projects and define ongoing action items necessary to complete individual tasks listed in these guidelines. The group finalized a set of goals to help shape future interactions and has published fact sheets that detail the missions, areas of expertise, and interests of each agency in relation to hydropower.

The FIHWG has laid out the following goals:

- **Goal 1**: Foster and maintain the interagency relationships established by the MOU by providing a forum where all participating agencies can exchange information, provide substantive input, and where possible, seek consensus on issues including the following:
  - Hydropower-related activities
  - MOU tasks
  - Perspectives on issues
  - Ongoing research
  - Potential legislation or funding sources proposed by Congress
  - Opportunities to maximize collaboration, increase efficiency, and avoid duplication of efforts
  - Environmental performance improvement, generation efficiency, and system reliability at hydropower facilities
  - New hydropower technologies and their potential benefits, risks, and impacts (including cumulative impacts).

  - **Goal 1 Outcomes**
    - Increase hydropower expertise among agencies to review hydropower project proposals, increase sustainable generation, and improve environmental performance.
    - Identify timelines for various MOU tasks/products and provide opportunities for input.
    - Recommend methods for incorporating FIHWG findings and information into the institutional practices of participating agencies.
    - Expand opportunities for interagency communication and networking.

- **Goal 2:** Enhance sustainable hydropower development by carrying out the following:
  - Identifying potential hydropower resources
  - Coordinating and sharing research
  - Exploring ways to reduce environmental impacts (including coordinated basin-scale development)
  - Communicating hydropower's value as a renewable resource.

  - **Goal 2 Outcomes**
    - Identify opportunities and assess alternative methods for implementing sustainable hydropower improvements (reduced environmental impacts and increased generation) at new and existing, federal and non-federal facilities.
    - Identify opportunities for collaboration and cooperative research by cataloging agencies' related ongoing and proposed research priorities and funding.
    - Evaluate the benefits and potentially pursue the development of a national standard or certification for low-impact hydropower development with input from key stakeholders, non-federal organizations, and existing certification groups.
    - Identify information resources that FIHWG agencies can use to communicate a consistent message regarding the value of hydropower to non-federal audiences.

- **Goal 3:** Create opportunities to better integrate and coordinate regulatory processes associated with non-federal hydropower development, including those at federal facilities, considering costs, time, complexity, and risk, while ensuring that all agencies' missions and responsibilities are met.
  - **Goal 3 Outcomes**
    - Develop an interagency "toolkit/handbook" on standard practices (proposed study protocols, example study templates, example license terms and conditions, etc.), and procedures to achieve improved consistency in information to provide greater clarity and predictability in the development of a sound evidentiary basis supporting improved and integrated project operations and mitigation.
    - Promote effective and efficient review of hydropower development permit applications by identifying the information or additional coordination (e.g., studies to establish baselines for impact assessment) needed by environmental resource agencies in support of FERC's efforts to communicate these needs to hydropower applicants.
    - Promote effective and efficient review of hydropower development permit applications by enhancing coordination and developing a template for communicating and emphasizing commitments to specific deadlines and time frames associated with federal agencies' statutory and/or regulatory responsibilities.
    - Investigate permitting processes to identify any redundancies and evaluate potential time savings and efficiency gains from the elimination of duplicative processes.

Timeline/Level of Effort: Hold quarterly meetings.

Scope/Location: FIGWG meetings will be held via teleconference or face-to-face group meetings.

Final Deliverable: Information sharing at quarterly meetings and an annual workshop focused on current, past, and future federal R&D efforts.

Lead Agency and Participating Agencies/Organizations: This group is led by DOE, Reclamation, and the Corps with the involvement of the 13 other federal entities listed below:

- BPA
- Bureau of Indian Affairs
- Bureau of Land Management
- EPA
- FERC
- Fish and Wildlife Service (FWS)
- U.S. Forest Service (USFS)
- National Park Service
- NOAA
- SEPA
- SWPA
- USGS
- WAPA

Participants are as follows:

DOE: WWPTO
Reclamation: Power Resources Office, Science and Technology Office
Corps: Corps Headquarters Operations

NEW: National Hydropower Vision

Needs Statement: Hydropower has been providing large amounts of renewable energy to the U.S. for generations, yet there is still potential for further development that can be sustainably pursued. Due to its regional nature, the hydropower community has specialized and fragmented, dealing effectively with localized issues, but without a national strategy. Therefore, DOE has undertaken the collaborative development of a long-term national Hydropower Vision for the benefit of the broad U.S. hydropower community.

Description: This landmark Hydropower Vision project will establish the analytical basis for an ambitious roadmap to usher in a new era of growth in sustainable domestic hydropower over the next half century. The objectives of the Hydropower Vision will be to:

- Analyze a range of aggressive but attainable industry growth scenarios
- Provide best available information relative to stakeholder interests
- Provide objective and relevant information for use by policy and decision makers.

Included in this effort will be:

- A close examination of the current state of the hydropower industry
- A discussion of the costs and benefits to the nation arising from additional hydropower
- A roadmap addressing the challenges to achieving higher levels of hydropower deployment within a sustainable national energy mix.

Work is organized by a core team at DOE, with topic-based task forces led by DOE's national laboratories and populated by the hydropower community generating the actual report. A senior peer review group made up of 16 senior industry members (including high-level officials from Corps and DOI) is providing sector-specific feedback throughout the project.

Timeline/Level of Effort: This activity will run from FY 2014 to 2016, with document completion and external review expected at the end of FY 2015, and DOE concurrence followed by public release in FY 2016.

Scope/Location: The report is national in scope, and as such, engagement of the hydropower community will include all regions, including Alaska and Hawaii.

Final Deliverable: The ultimate deliverable is a DOE report several hundred pages in length, covering the state of the hydropower industry, an analysis of likely futures, and a roadmap.

Lead Agency and Participating Agencies/Organizations: DOE will be leading this effort with task forces captained by its national laboratories. Reclamation, as well as other parts of DOI and Corps will provide significant support as task force members, senior peer review group members, and reviewers.

DOE:	WWPTO, ORNL, National Renewable Energy Laboratory (NREL), PNNL, ANL, BPA, SWPA, SEPA, WAPA
Reclamation:	Power Resources Office, Technical Services Center, Research and Development Office
Corps:	Corps Headquarters Operations, HDC's HAC

NEW: Hydropower R&D Dissemination Database

Needs Statement: A substantial amount of R&D of new hydropower technologies and techniques has been carried out by the MOU agencies in past years, and more new work is planned for the future. However, no common database exists for the MOU agencies to share information with the public or among themselves. Without this information, the MOU agencies run the risk of duplicating efforts or failing to leverage existing research, and non-federal groups may not be adequately able to leverage federal investments for the public good. Therefore, a common database will be developed to allow the agencies to upload and share information as it relates to the MOU.

Description: The MOU agencies will identify an appropriate platform for the database (e.g., website, SharePoint, cloud, etc.) and determine how best to maintain the database. The MOU agencies will then be responsible for uploading their information and making updates as necessary. The MOU agencies may need to discuss how best to protect and/or restrict access to any sensitive information.

Timeline/Level of Effort: Efforts to develop the database will begin in FY 2016.

Scope/Location: The database will initially cover Corps and Reclamation facilities across the nation.

Final Deliverable: Creation of a database that will allow the MOU agencies to upload and publicly share information as it relates to the MOU.

Lead Agency and Participating Agencies/Organizations: DOE will take lead in developing the database, which will host information from other federal agencies, including but not limited to DOE, Corps, Reclamation, WAPA, BPA, and TVA.

Participants are as follows:

DOE:	WWPTO
Reclamation:	Power Resources Office/Science and Technology Office
Corps:	HDC

NEW: Providing Tools and Increasing Coordination to Improve Permitting Efficiency

Needs Statement: Complex federal permitting processes for hydropower that involve multiple federal agencies can often appear unclear and burdensome to developers. It can also be difficult to coordinate the flow of information and associated timelines for concurrent phases of review and consultation between developers and agencies. Additionally, recent hydropower resource assessments have identified significant opportunities to develop additional hydropower on federal infrastructure or on federal lands; external stakeholders have made it clear that improving coordination of permitting processes amongst federal agencies will be a critical step to realizing these opportunities.

Description: In alignment with Executive Order 13604 and the President's Climate Action Plan (which both focus on improving processes for federal permitting and review), the Renewable Power Technology Offices within EERE have worked together to develop the first components of a Regulatory and Permitting Information Desktop (RAPID) Toolkit, the development and use of which would occur in partnership with other federal agencies. The ultimate goal of this initiative is to develop a comprehensive one-stop permitting resource and application toolkit that will facilitate federal-level permitting, and federal-state interactions for regulating renewable energy (and in this case, hydropower) projects. The system could incorporate abilities to track permitting processes for individual projects from beginning to end across agencies and capture analytical data to inform future improvement efforts. The toolkit could also provide information at each step on referenced regulations, agency processes, similar National Environmental Policy Act analyses, rules/guidelines, points of contact, timelines, etc. Initial experience with a similar pilot effort initiated by EERE's Geothermal Technologies Office has been positive. The Bureau of Land Management and USFS field offices are utilizing it as a training manual and step-by-step handbook to administer their process, state officials in California have identified specific process improvements that are being implemented, and geothermal project developers report an increasing degree of standardization and clarity in the permitting process. The initial goals for the hydropower-specific RAPID Toolkit are to ensure that standardized, accurate, and easily digestible information exists for federal permitting processes for hydropower technologies, while working closely with federal agencies to ensure that new resources are useful and being utilized by staff. If successful, the toolkit could potentially be expanded to allow developers to submit permitting information to multiple federal agencies or offices simultaneously. It could also allow for applications to be tracked through the permitting process for greater transparency and discover where further improvements could be made.

By coordinating this type of new initiative through the FIHWG (which FERC and other federal resource agencies participate in), this effort will advance the group's mutual goal to "*Create opportunities to better integrate and coordinate regulatory processes associated with non-Federal hydropower development, including those at Federal facilities, considering costs, time, complexity, and risk, while ensuring that all agencies' missions and responsibilities are met.*" This initiative will help to build upon the existing relationships established under the FIHWG to increase coordination and dialogue between all necessary agencies and individuals, with the ultimate goal of reducing potential regulatory inefficiencies and increasing hydropower deployment. Ultimately, this will save time and money for both developers and federal agencies.

Timeline/Level of Effort: This activity would be planned for initiation in FY 2015, with the first stages of development occurring throughout FY 2015 and FY 2016. Depending on the level of complexity scoped for the initial development of the system, considerable amounts of staff time might be necessary from multiple federal agencies for review and feedback in the latter half of FY 2015 and throughout FY 2016.

Deliverable: The goal for the first phase of this effort would be to have a publicly accessible beta version of an online information tool available during FY 2016.

Lead Agency and Participating Agencies/Organizations: DOE and its national laboratories will take the lead in the initial development of this system and consultation/refinement with other federal agencies and stakeholders, while federal agencies engaged in hydropower regulation and permitting (including Corps, Reclamation, FERC, FWS, USFS, National Marine Fisheries Service, and others) would ultimately be responsible for participating in demonstration and implementation.

Participants are as follows:

DOE:	WWPTO, NREL
Reclamation:	Power Resources Office
Corps:	Corps Headquarters Operations

MOVING FORWARD

The MOU agencies have made significant progress since 2010 and intend to continue working together to improve and expand hydropower generation in the United States. Electricity from hydropower still represents the majority of renewable energy capacity and generation across the country, and the flexibility and grid services provided by hydropower can be extremely useful for the integration of other renewables as they continue to expand.[6] Although hydropower is an extremely reliable and long-term resource, analysis shows that more than half of all hydropower turbines in the U.S. are more than 50 years old (35% are more than 75 years old). This presents both a challenge and an opportunity in the need to innovate and modernize U.S. facilities so that the nation can continue to rely on its hydropower resources for cost-effective and renewable electricity into the future. Non-hydropower federal water infrastructure (namely, non-powered dams and canals) also present significant opportunities for further hydropower development, provided that processes are in place to allow for efficient and timely development. There are also numerous opportunities to develop new generation through research and demonstration of advanced technologies, which can further reduce the cost of energy for all types of new hydropower and improve sustainability and environmental performance, which will have benefits both domestically and internationally. As the hydropower industry continues to grow in countries outside the U.S., there is a unique opportunity to export new sustainable technologies and techniques to partner nations, and help advance the spread of clean energy around the world. Corps, DOE, and Reclamation will continue to work through their partnership and expect to make even more progress over the coming years.

---

[6] 2010 State Renewable Electricity Profiles: http://www.eia.gov/renewable/state/.

www.ingramcontent.com/pod-product-compliance
Lightning Source LLC
Chambersburg PA
CBHW080622180526
45168CB00007B/3017